**BODY SYSTEMS**

# Moving

## Jackie Hardie & Angela Royston

Heinemann

First published in Great Britain by Heinemann Library
Halley Court, Jordan Hill, Oxford OX2 8EJ
a division of Reed Educational and Professional Publishing Ltd.

OXFORD FLORENCE PRAGUE MADRID ATHENS MELBOURNE
AUCKLAND KUALA LUMPUR SINGAPORE TOKYO IBADAN
NAIROBI KAMPALA JOHANNESBURG GABORONE
PORTSMOUTH NH (USA) CHICAGO MEXICO CITY SAO PAULO

Designed by Inklines and Small House Design
Illustrations by Catherine Ward, except: p.6, p.11 (right), p.12, p.20,
p.22 & p.26 by Frank Kennard.

Printed in Great Britain by Bath Press Colourbooks, Glasgow
Originated in Great Britain by Dot Gradations, Wickford

01 00 99 98 97
10 9 8 7 6 5 4 3 2 1

ISBN 0 431 06207 2

This title is also available in a hardback library edition (ISBN 0 431 06206 4).

**British Library Cataloguing in Publication Data**
Hardie, Jackie, 1944 –
    Moving. – (Body systems)
    1. Human mechanics  – Juvenile literature 2. Kinesiology – Juvenile literature
    I. Title
    612.3

### Acknowledgements
The Publishers would like to thank the following for permission to reproduce
photographs:
Action-Plus Photographic: p.5, p.14, p.28; Allsport USA: p.13; Hulton Deutsch
Collection: p.27; NHPA/Martin Harvey: p.25; Oxford Scientific Films: p.10;
Science Photo Library: p.3, p.7 (both), p.8, p.9, p.19, p.20, p.21, p.23 (both), p.29;
Tony Stone Images: p.4, p.24; Zefa: p.17.

Cover photograph: Trevor Clifford.

Our thanks to Yvonne Hewson and Dr Kath Hadfield for their comments in
the preparation of this book.

Every effort has been made to contact copyright holders of any material
reproduced in this book. Any omissions will be rectified in subsequent printings
if notice is given to the Publisher.

# Contents

# You need to move

Almost everything you do involves movement, except perhaps thinking. But even when you are thinking, you may scratch your head, tap your pencil or pace up and down! You need to move, not only to get about, but to eat, get dressed, read a book or do whatever you want to do. Moving means using your muscles and joints to move your bones.

◀ *This soldier on guard duty outside Buckingham Palace, London, is trying his best not to move. He seems not to be moving at all, but he still has to breathe and blink his eyes.*

## Bones and joints

Without bones you would be floppy and shapeless, like a jellyfish stranded on a beach. Bones give you your shape, but bones cannot bend. If you had only one bone in each leg and arm, you would move very stiffly. You can bend and move your body because you have many bones.

The places where bones meet are called joints. There are joints at your knees, elbows, ankles, wrists, fingers and toes. There are also many joints along your backbone which allow you to bend and twist. Joints are moved by muscles.

## Muscles

There are three kinds of muscle in your body. Between a third and a half of your weight is due to **skeletal muscles**, the meaty kind of muscles which are joined to and move the bones. If you lift your leg straight up you can feel this type of muscle in the top of your leg stiffening and working.

Another kind of muscle, called **smooth muscle**, is found in the walls of the **intestines**. It moves food through the **digestive system**. Your **heart** is made of the third kind of muscle, called **cardiac muscle**, which pumps blood around your body. Blood carries all the fuel and **oxygen** the bones and muscles need to work, grow and repair themselves.

► *Playing baseball involves moving your arms, legs and most of the bones and muscles in your body.*

## Did you know?

*All animals move, but not all animals have bones. Insects have a hard covering outside their bodies which gives them their shape. The muscles which move their limbs are attached to this hard outer skeleton (called an exoskeleton). Worms and jellyfish have no hard skeleton at all.*

# Your skeleton

Babies have more bones than adults. As a newborn baby you had over 300 separate bones, but as you grew, some bones joined together, or became fused. By the time you are an adult you will have 206 bones. Each one is joined to another one and together they make up your skeleton. The skeleton not only gives your body its shape, it supports you and protects **vital organs**, such as your **brain**, **heart** and **lungs**.

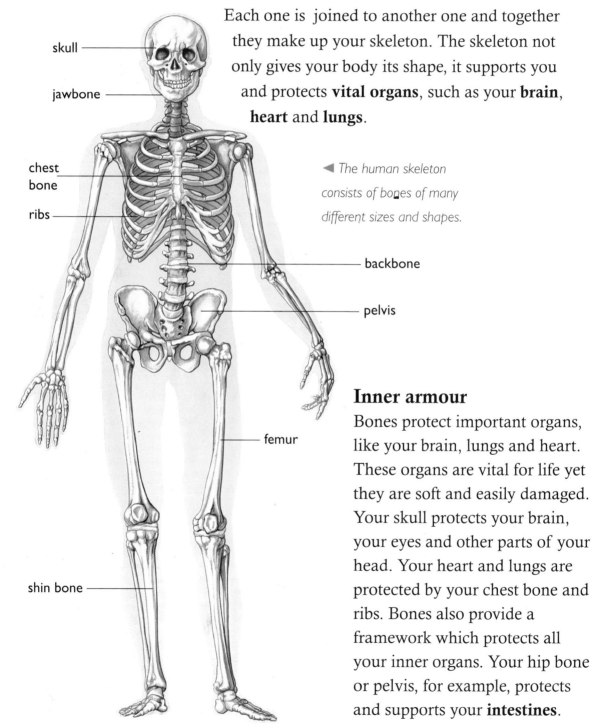

skull

jawbone

chest bone

ribs

◀ *The human skeleton consists of bones of many different sizes and shapes.*

backbone

pelvis

femur

shin bone

## Inner armour

Bones protect important organs, like your brain, lungs and heart. These organs are vital for life yet they are soft and easily damaged. Your skull protects your brain, your eyes and other parts of your head. Your heart and lungs are protected by your chest bone and ribs. Bones also provide a framework which protects all your inner organs. Your hip bone or pelvis, for example, protects and supports your **intestines**.

## A tough job

Bones are as hard as concrete and longer lasting. They are made of a very strong substance called **collagen** mixed with **calcium** and **phosphates**. Bones have to be tough, not only to protect your insides from damage, but also to carry the weight of your body. They even take the strain of carrying extra weights, such as a heavy bag, and the bones of your legs can withstand the jolt when you jump off a wall.

Children's bones can squash and bend more easily than adults' bones. Newborn babies have very soft, bendy bones to help them squeeze out of their mother's body. Their bones harden up as they grow.

▲ This ancient skeleton is between 30 and 60 thousand years old. It shows that bones and teeth are the toughest parts of the body.

▼ The smallest bone in the body is the **stirrup**. It is one of three bones in the ear. In this photo it is about 10 times larger than life size.

### Did you know?

The largest bone in your body is the thigh bone, also called the femur. It gives you about a third of your total height. The smallest bone, the stirrup, is in your ear. It is only about 3 mm long, less than the length of a grain of rice. The movements, or vibrations, of this and two other tiny bones carry sound waves to your inner ear.

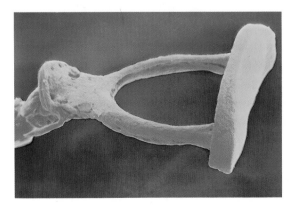

# Inside a bone

Bones are not only strong, they are light as well. Heavy bones would slow us down and tire us out. Bones are light because they are not solid. The outer **compact bone** is thick and hard, but the inside is a mesh of thin bone with spaces in between. The mesh is called **spongy bone**, but it is almost as strong as steel. In many bones the spaces are filled with **bone marrow**. The whole bone is fed by masses of tiny blood vessels.

## Living bone

When you think of bones you probably think of dried up ones like the dead animal bones you might find. But the bones in your body are very much alive, and bone **cells** constantly renew themselves. Bones are surrounded by a **membrane** of blood vessels and **nerves**, which supply the bone with the food and **oxygen** that the living cells need.

▲ *Spongy bone magnified 50 times.*

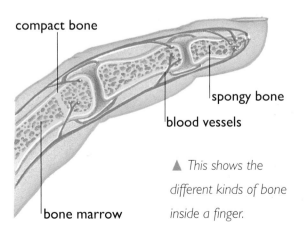

compact bone

spongy bone

blood vessels

bone marrow

▲ *This shows the different kinds of bone inside a finger.*

## Bone marrow

Bone marrow is a soft, red jelly which fills the spaces in some spongy bone. It also fills the hollow tube at the centre of the long bones of the legs and arms. Bone marrow makes red and white blood cells. Red blood cells carry oxygen from the **lungs** around the body. White blood cells fight and kill the **bacteria** and **viruses** which cause disease.

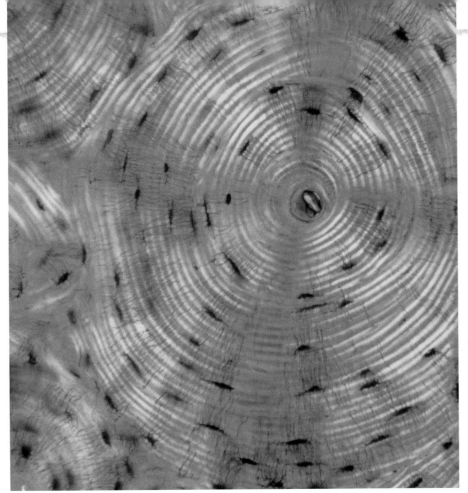

◄ Compact bone, which is thick and hard, magnified about 150 times.

## Cartilage

When bones grow, they first form **cartilage**. Cartilage is made up of the **protein** called **collagen**. If you feel your ear you will find it is rubbery. Ears are made of cartilage, and so is the end of your nose. Most cartilage gradually changes into hard bone.

The cartilage in your ears and nose never becomes hard bone. Even when your body bones have stopped growing and hardening, the ends remain covered with cartilage to stop them grinding together and damaging each other at the joints.

### Did you know?

Red blood cells do not last very long, only about three months, so your bones are busy making new cells all the time. Bone marrow makes new blood cells at an extraordinary rate – 2 million red blood cells every every day!

# Skull and backbone

Your skull protects your **brain**, eyes and ears. It is made of several flat **compact bones** which are fused together. Only the lower jaw can move. Your backbone, or spine, protects your **spinal cord**, the main **nerves** which connect the brain and the body. It consists of 33 separate bones, called **vertebrae**, stacked one on top of the other.

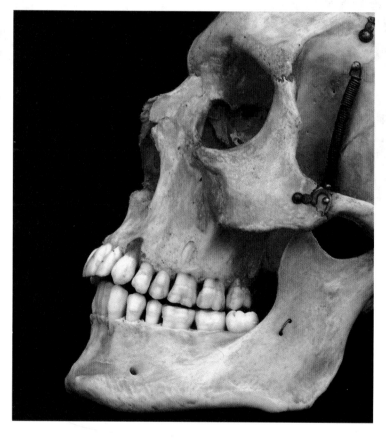

◄ *Some of the bones of the skull. In this picture the spring fitted to the skull shows you where the jaw hinge is. The nose looks short because it does not include any cartilage.*

## Your skull

The main part of your skull is the rounded part at the top called the **cranium**. It is made of eight bony plates fixed tightly together. Inside, the brain floats in a cushion of liquid. The skull has only a few gaps in it.

These gaps are holes for your eyes, ears, nose and spinal cord. The lower jaw is hinged so you can move it to eat and talk. Teeth are even harder and tougher than bone. They are set in the strong bones of the jaws.

## Soft spots

A baby's skull is not hard and the bones are not fused together. The bones are soft and there are several soft spots called **fontanelles** made of **membrane**. They allow the head to change shape as the baby squeezes out of the mother's body. Once the baby is about a year old, the skull has hardened. The fontanelles have changed first to **cartilage** and then to bone.

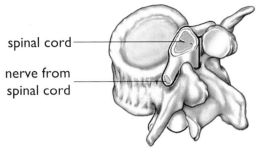

▲ One of the knobbly bones called vertebrae which make up the spine.

## Your spine

The spine is very strong. It not only supports the weight of your head, but all the main bones of your body are directly or indirectly joined to it. Each vertebra is shaped like a knobbly ring. If you feel your own back, you will feel some of the bumps. The spinal cord runs down the centre of the vertebrae. The ribs are attached to the side of the spine and these help to protect the **heart** and **lungs**. There is a cushion of cartilage between each of the vertebrae. When you bend, this cartilage, or disc, is squashed to let the bones move easily.

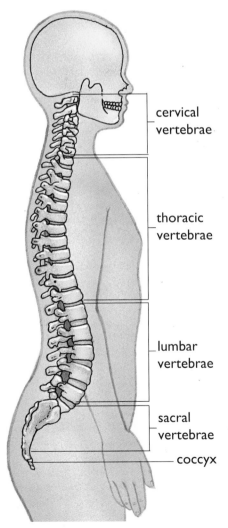

▲ The spine curves slightly into the shape of an S, which allows you to bend more easily.

### Did you know?

You are about a centimetre taller first thing in the morning than when you go to bed. During the day, the cartilage between the bones of the spine becomes slightly squashed. During the night it springs back into shape.

# Arms and legs

You have long legs that allow you to get about quickly and easily, and long arms that help you reach for things and throw with some force. Each leg and arm has three long bones. They are stronger than most other bones, particularly at the ends where the **spongy bone** is packed tightly together. The centre is hollow and filled with **bone marrow** to keep the bone light.

## Shoulders and hips

Your legs and arms are attached to the rest of your body at your hips and shoulders. Your shoulder blade and collar bone join your arms to your spine and chest bone. Your shoulder blade is thick and strong to take the weight of your arms. Your collar bone is more delicate. Your hip bone, or pelvis, joins your legs to your spine and it, too, is very strong.

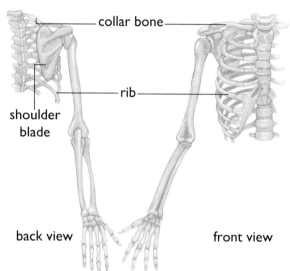

▲ The arm bones are supported by the shoulder blade and collar bone.

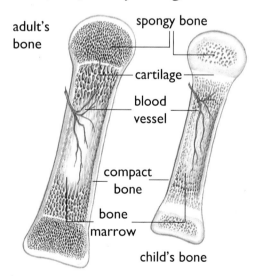

▲ Long bones are strong to carry weight. Children's bones have more cartilage at the ends, which gradually hardens into bone.

## Long bones

The long bones grow by forming new **cartilage** near the ends, which gradually changes into hard bone. When the bones first formed in the womb, they formed as cartilage. Even when the baby is born, many of his or her bones are still soft. The older you get, the harder your bones become. For very old people, this can become a problem. If they fall, their bones are so brittle they break easily.

◄ Collar bones are easily broken. American footballers protect their shoulders with padding. They also wear helmets to protect their skulls.

## Did you know?

All mammals have the same kinds of bones as we do. They vary in shape, particularly in the limbs, but amazingly a seal's flipper is shaped by the same bones as our feet and legs, a bat's wings are made by flaps of skin stretched between its very long fingers, and what you might think is a dog's knee is really its ankle.

# Joints

We can only bend and move our bodies at the joints – the places where two or more bones meet. Different kinds of joints allow different kinds of movement. Some joints do not move at all – the bones at the bottom of your spine, for example, are fused together and cannot move. Each kind of joint is constructed so that it allows a particular kind of movement, but makes other kinds of movement impossible.

## Hinge joints

However hard you try, you can only move your finger at the knuckle in two directions – up and down. Fingers and toes have hinge joints, so called because they work like the hinge on a door. Knees and elbows are hinge joints, too. This makes your legs and arms more stable. Your legs would tend to give way under you if they could twist at the knee.

## Ball and socket joints

Hips and shoulders are ball and socket joints. The upper end of the thigh is rounded and fits into a deep cup, or socket, in the hip bone. Similarly, the rounded top of the arm fits into the shoulder blade. You can not only swing your arm or leg, you can move them around in circles.

◀ *The knee can only move backwards and forwards. It is a hinge joint.*

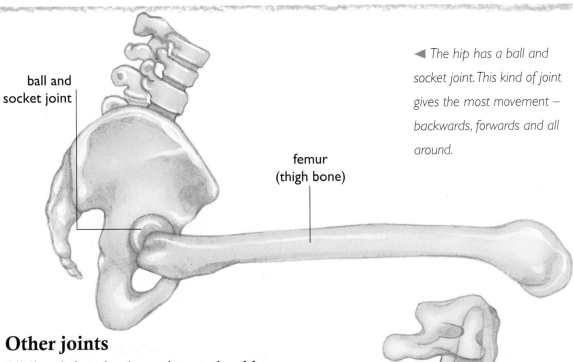

ball and
socket joint

femur
(thigh bone)

## Other joints

Sliding joints in the wrists and ankles allow you to move your hands and feet from side to side as well as up and down. Although each of your fingers can only move up and down, your thumb can move all around. That is because it has a saddle joint, so called because it looks like a horse's saddle. Your neck needs to be strong to hold up your head. It has a pivot joint which allows you to turn your head as well as nod it up and down.

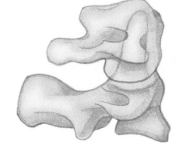

▲ The neck has a pivot joint.

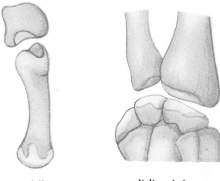

saddle joint          sliding joint

▲ Wrists and ankles have special joints to allow a wide range of movement.

### Did you know?

Many elderly people have an artificial hip. The hip joint, however, is the strongest joint in the body. It took scientists some time to come up with an artificial joint which was strong enough to take the strain. Modern hip replacements have a steel ball which moves in a polythene socket or cup.

# Inside a joint

Joints have to allow the bones to move smoothly, but they also have to be strong. **Ligaments** are tough bands of gristle which bind the bones together. If the bones rubbed against each other when they moved, they would grind, creak and soon wear out. The ends of the bones are protected and cushioned by slippery **cartilage** and some joints are oiled by a liquid called **synovial fluid**.

## Ligaments

Ligaments are strong but slightly stretchy. They stop the bones from moving too far and so damaging the joints. Gymnasts and dancers do special exercises to stretch the ligaments so that they can perform a greater range of movements. If they overdo it, however, they may damage their joints, making them stiff and painful much sooner than other people's.

Tennis players and footballers, on the other hand, need to have very strong joints. They have to run, stop and change direction fast. This puts a lot of strain on their knees and many players wear strong elastic bandages to give their knee joints extra support when they are playing.

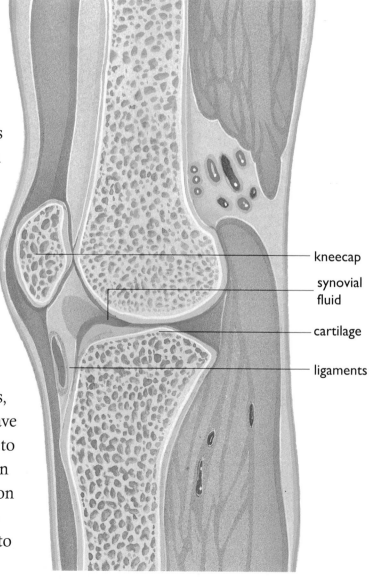

kneecap

synovial fluid

cartilage

ligaments

▲ The inside of a knee joint.

## Arthritis

Cartilage is almost completely smooth, but as people get older, the cartilage in their joints wears down. This is called arthritis. It causes the bones to become rough and the joints very painful. The hips and the joints in the spine are usually the worst affected as they carry the most weight.

If the arthritis is very painful, a person may be given a **hip replacement** operation or the joints in the back may be fused. Another form of arthritis causes the joints in the hands and wrists to become swollen and painful, and these may become difficult to move.

▲ Gymnasts can move their joints further than most people.

### Did you know?
There is no such thing as a 'double joint'. Some people do have extra long ligaments, though, so they can move their joints in very unusual ways. This is what is called a 'double joint'

# The muscles

Bones provide a framework for your body, but the muscles which cover the bones give the body its final shape. Every movement you make is controlled by many different muscles working together. Walking uses over 200 muscles, whilst even a smile uses 15. Muscles work by becoming shorter (contracting).

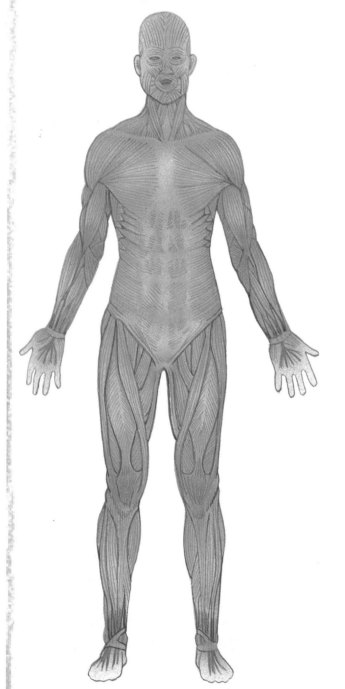

## Moving bones

One end of a muscle is anchored to the bone it covers, but the other end is attached to a **tendon** which joins the muscle to another bone. The tendon usually crosses a joint. When a muscle contracts, it pulls on the tendon to move the bone at the joint. The muscles in your thigh pull on the lower leg, so you can bend and straighten your knee. If you sit and swing one foot back and forth, you can feel the tendons working at the back of your knee.

The muscles in your lower leg move your feet, and the stomach muscles lift your legs and allow you to bend and straighten. The large flat muscles across the chest and back are attached to the upper arms.

◄ *The main muscles of the body. Your body is symmetrical, so you have the same muscles on the right and left sides of your body.*

## Other kinds of muscle

The muscles which cover your body are called **skeletal muscles**. They do not all move bones. The **diaphragm**, for example, has a strip of muscles around its edge and it helps you breathe. Your tongue is a muscle, and muscles control the blinking of your eyelids. Some people can even contract the muscles attached to their ears!

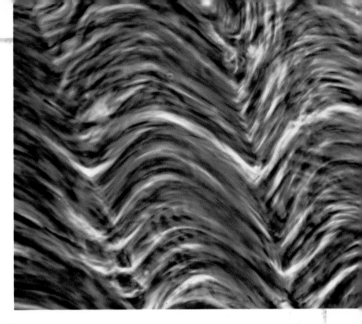

▲ *Smooth muscles in the walls of the digestive system move food through the body. They work by squeezing and pushing the food along.*

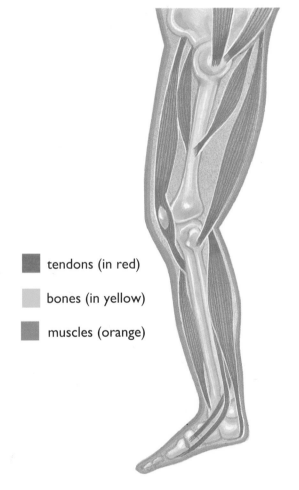

tendons (in red)

bones (in yellow)

muscles (orange)

▲ *Bones and muscles work together to produce movement. The muscles are attached to the bone by tendons. The tendons from the muscles in the thigh move the lower leg by pulling on it.*

A different kind of muscle, **smooth muscle**, works deep inside your body, moving food through the **digestive system** and blood through the **veins** and **arteries**. Smooth muscles work by squeezing the tube behind the food or blood, rather like you squeeze a toothpaste tube. The **heart** is made of **cardiac muscle**. When it contracts, blood is pumped into the **lungs** and around the body.

### Did you know?
You have about 650 muscles in your body and together they make up about four-tenths of your weight. The largest muscle is the gluteus maximus, also called the buttock. It raises your thigh and gives you something comfortable to sit on. You can feel it working when you stand up.

# Inside a muscle

Skeletal muscles are made of millions of tiny strands called **muscle fibres**. Each strand consists of even thinner fibres called filaments. You can see the fibres if you look at raw meat. The fibres are held together in a bundle by a kind of thin skin called a **membrane**. A slightly thicker membrane holds the bundles together to make the muscle.

## Exercise

**Nerves** carry electrical signals from the **brain** to make the muscle fibres contract. The more you work a muscle, the bigger and stronger it becomes. It does not get bigger by adding more fibres, but by working better. Even wrestlers with huge muscles have no more muscle fibres than the day they were born!

If you don't use a muscle, it becomes thin and weak. If you break an arm or leg, a doctor covers it with a plaster cast until the bone heals. This stops the joint moving, so the muscles which normally work across the joint weaken. The joint needs exercising to strengthen the muscles when the plaster is removed.

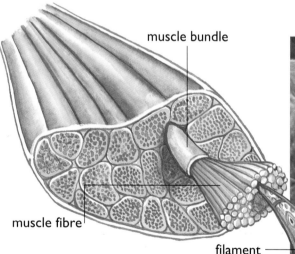

muscle bundle

muscle fibre

filament

▲ *A muscle is made up of fibres arranged in bundles. Each fibre is made of tiny filaments.*

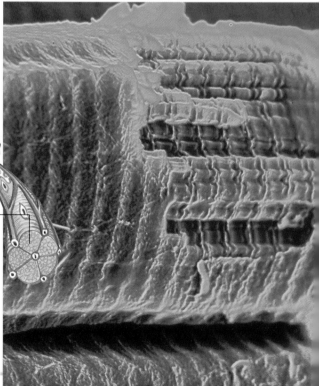

▶ *This muscle fibre has been enlarged over 1000 times.*

## Muscle fuel

Muscles need **oxygen** and energy to work. Oxygen comes from the air we breathe and energy comes from food. The **digestive system** breaks food into different substances including **glucose**, which is a kind of sugar.

Both oxygen and glucose are carried to the muscles by the blood. As muscles use the glucose, energy is released. Some of this energy is in the form of heat, which is why you get hot when you exercise.

◀ *Exercise makes muscles bigger and stronger.*

### Did you know?

You get a 'stitch' when your muscles run out of oxygen. If you exercise very hard, the blood cannot always get enough oxygen to the muscles. This means cells do not burn glucose completely and produce a substance called lactic acid. Too much lactic acid causes the painful stitch.

# How muscles work

Muscles cannot push, they can only pull. Even when you push against a wall, each muscle in your body is working by pulling. When the muscles relax, they stop pulling. There is more than one muscle that works each joint. The muscle that bends your arm cannot straighten it alone.

## Working in pairs

Most muscles, especially the ones that move your bones, work in pairs. As one muscle contracts and gets shorter, another relaxes and returns to normal length. The **biceps**, for example, contracts to bend your arm, while the **triceps** at the back of your arm relaxes. If you simply relax the biceps, your arm will stay bent. To straighten it again, you have to contract the triceps. The action of one muscle is opposed by the other.

Pairs of muscles work all the time to help you keep your balance, to move your head, back, arms and legs. Even the **smooth muscles** in your **digestive system** work against each other. While one muscle squeezes the tubes involved in digestion, another straightens them out again. This pushes the food along.

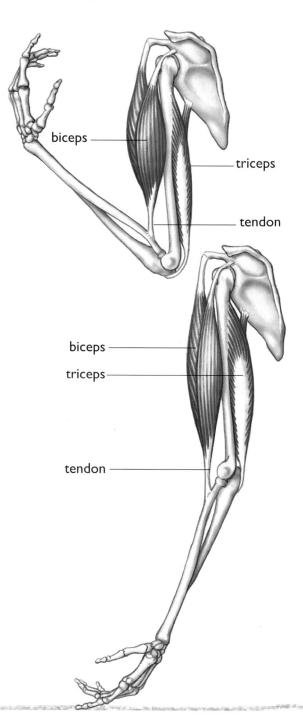

▶ *The biceps bends the arm. The triceps straightens it.*

## Your face muscles

Your face has many muscles which control your jaws, eyes and skin. You use more than 30 facial muscles to make a whole range of different expressions – frowning, smiling, scowling and so on. Some facial muscles are attached to the skull bones, others are attached to the skin.

These muscles work by contracting and relaxing. The iris is a circular muscle. It controls how much light enters the eye through the pupil. In bright light, the pupil is small as the iris has contracted. In dim light it relaxes, to make the pupil larger and let in more light.

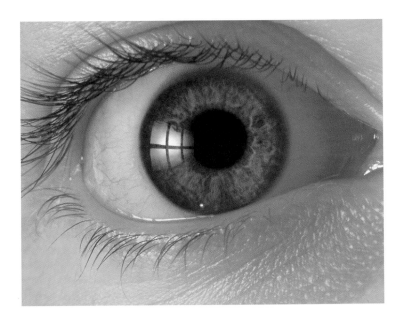

◀ *The iris, which gives your eyes their colour, controls how much light enters the eye through the pupil. In bright light the pupil is smaller to let in less light (top). In dim light the pupil is enlarged to let in more light (bottom).*

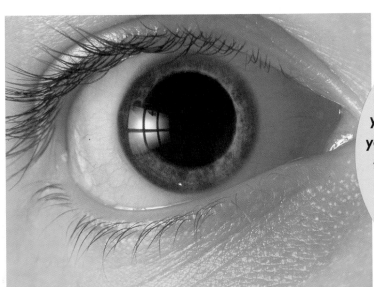

**Did you know?**

The fastest-acting muscle in your body is the one that makes you blink. You can blink up to five times a second. However, this is slow compared to the speed at which some insects can flap their wings – 1000 times a second.

# Hands and feet

The human hand is an amazing machine. It can do precise and difficult tasks, it can move quickly, and it has a powerful grip. No machine can copy all the things your hands can do. Your thumb is the key to your hand's success. Your thumb's lowest joint is a saddle joint, which allows it to move in all directions and to touch each of your other fingers.

## Opposable thumb

Squirrels hold nuts in their hands to nibble them, but they have to use both hands to hold the nut. Humans and some apes, however, are able to pick things up with just their thumb and one or two fingers. Our thumb is said to be opposable, and it allows us to hold and move anything from a pen or paintbrush to a hammer or steering wheel.

▲ The human hand is designed to do many small, complicated movements.

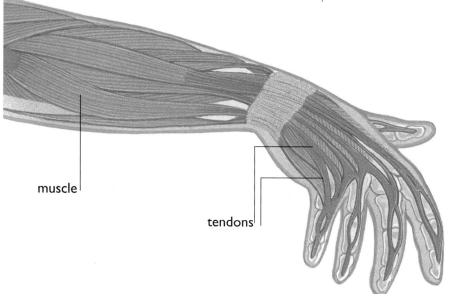

muscle

tendons

◀ Many of the muscles which move the fingers are in the lower arm and are connected to the fingers by long tendons.

## Long tendons

The skin on the tips of your fingers and thumbs is very sensitive. You can feel small grains of dust and even pick up a single crystal of salt. Some of the muscles which let you do this are in your hand and some in your lower arm. Look at the back of your hand while you move your fingers. Can you see the **tendons** from the lower arm moving just beneath your skin?

Feet are designed in a similar way to hands. A tendon called the Achilles tendon crosses the back of your ankle. It joins the muscles at the back of your calf to your heel. When you walk, you press down on your heel and use your toes to lever yourself along. Athletes sometimes injure their Achilles tendon when training hard.

▲ Some kinds of ape, like this young orang-utan, can grip with their toes and fingers just as we grip with our hands. They can use hands and feet to hold food and to grip the branches of trees.

### Did you know?
Half of all your bones are in your wrists, hands, ankles and feet. Each hand and foot has 26 bones. The bones in your wrists and ankles are small and squarish. Their sliding joints let you move your hands and feet from side to side, up and down and round in circles.

# Breaks and sprains

Your body is very strong and stands up to a lot of wear and tear, but sometimes a fall or an accident can damage the bones, joints or muscles. Usually the damage repairs itself, but a broken bone may need a plaster cast to support it while it mends. An elastic bandage is usually enough to support a sprained joint. Any serious injury should be seen by a doctor to find out how bad the damage is.

## Broken bones

A bone may be cracked, chipped or broken. Only an **X-ray** will show the actual damage. A doctor checks the X-ray and makes sure that the bone is in the correct position before a plaster bandage is wrapped around it. This supports the broken bone and stops it moving while it heals itself. As soon as a bone is injured, blood and **bone marrow** begin to ooze out and fill the gap. Slowly new bone grows and hardens.

## Sprains

Joints are surrounded by **membranes** and other tissues which protect them. If a joint is twisted out of position, the **ligaments** help to stop the joint from being damaged. Only a really violent movement will tear the ligaments. A sprain occurs whenever any of the tissue around the joint is damaged.

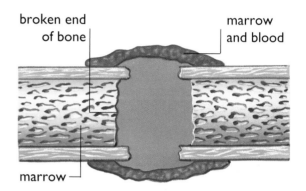

broken end of bone — marrow and blood

marrow

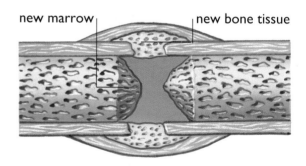

new marrow — new bone tissue

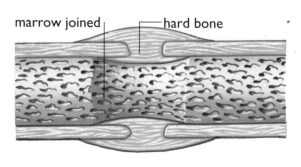

marrow joined — hard bone

▲ How a broken bone mends itself.

Your ankle and wrist are the easiest joints to sprain. A damaged joint may swell up and will certainly be very painful indeed.

Putting something cold, like a packet of frozen peas, on the joint helps to reduce the swelling. An elastic bandage gives support while it heals.

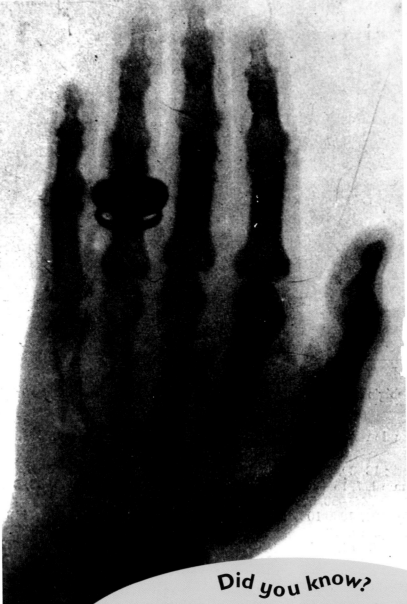

◀ *This is the first X-ray that was ever taken of a human hand. It belonged to the wife of Wilhelm Roentgen, the scientist who discovered X-rays in 1895.*

**Did you know?**

When X-rays were first discovered, people were amazed that photographs could be taken of the inside of their bodies. Some people were also worried that snoopers could use them to see through their clothes, so one company began to sell X-ray-proof underwear!

# Fit and healthy

If you want to keep fit and look good, you need to exercise regularly, eat healthy food and get enough sleep at night.

## Exercise

You can strengthen particular muscles by doing special exercises, but there are many ways of exercising your whole body. Swimming, football and other sports exercise your muscles and make your **lungs** and **heart** work better too. Lots of everyday tasks involve bending and stretching in ways that also help keep you fit.

## Body-building food

You need a balanced diet of different kinds of food to stay healthy. Fish, cheese, eggs and meat are particularly rich in **protein**. Proteins are complicated substances that the body needs in order to grow. They renew the millions of different **cells** that make up the body and repair damaged cells. If you want big, strong muscles, you should certainly eat lots of protein.

**Calcium** and **phosphates** are the **minerals** which make bones strong. While your bones are still growing, your body needs a constant supply of them. Milk and cheese are rich in calcium, but many other foods contain some calcium too.

◀ *The correct way to lift any heavy weight is to bend the knees and keep the back straight, just as this weight-lifter is doing.*

## Sleep

Sleep allows your body to rest. People who have not slept at all for just two days quickly become bad tempered and cannot concentrate on even the simplest tasks. When you are asleep your whole body has the chance to slow down and unwind.

Your muscles relax, your **heart-beat** and breathing slow down. The body's energy can then be used in other ways – to replace worn out or damaged cells and to fight illness. You actually grow more when you are asleep than when you are awake.

◀ Astronauts have to exercise hard to keep fit while they are weightless in space.

### Did you know?

Astronauts working in space are weightless. Their muscles scarcely have to work to support and move their bodies. This sounds very relaxing, but unused muscles soon become weak. In the 1960s it was found that astronauts' bones also became lighter when they worked in space. Today astronauts in space have to exercise to keep their bones and muscles healthy.

# Glossary

**Arteries**  Tubes which carry blood containing oxygen and dissolved food to the body cells.

**Bacteria**  Single living cells that we can only see through a microscope. There are millions of bacteria in the air all around us and in all parts of our bodies. Most are harmless, but some can cause diseases.

**Biceps**  One of the muscles in the upper arm. It causes the arm to bend when it contracts.

**Bone marrow**  A soft, red, jelly-like substance which fills the spaces in the spongy bone at the centre of some bones. Bone marrow makes white and red blood cells.

**Brain**  The soft, grey organ found inside the skull, which coordinates every single thing that we do, including breathing, eating, moving, thinking, learning and enjoying ourselves. The brain sends and receives messages as electrical signals to and from all parts of the body via nerves.

**Calcium**  A mineral which our body needs to keep bones strong and healthy. It is found in milk and cheese, and in many other foods.

**Cardiac muscle**  The walls of the heart are made from cardiac muscle, which is very strong, and never gets tired like ordinary muscle. This is very important, as the heart must keep beating to pump blood around our bodies to keep us alive.

**Cartilage**  A firm, but smooth and elastic substance found inside ear flaps and at the end of the nose. It cushions bones where they meet at joints, and stops them rubbing together. When new bone first forms it starts out as cartilage, but gradually hardens into bone.

**Cells**  The smallest living unit. Each part of the body is built up of a different kind of cell. Each cell has a nucleus which controls what it does and each cell is surrounded by a cell wall.

**Collagen**  A very strong protein which is one of the substances needed to make healthy bones.

**Compact bone**  The hard outer layer of bone that surrounds the lighter, spongy bone inside. Compact bone has tiny holes in it through which blood vessels carry blood in and out of the bone.

**Cranium**  The top part of your skull which helps to protect the brain by forming a case around it.

**Diaphragm**  The sheet of tissue beneath your lungs which is moved by muscles to control breathing. The muscles contract and relax to make you breathe air in and out.

**Digestive system**  The parts of the body which are used to digest food.

**Femur**  The long bone in the thigh.

**Fontanelles**  Soft spots on the top of a baby's skull, where the bones have not yet fused together. As the baby grows older, the bones will join up to form a hard skull.

**Glucose**  A kind of sugar which the body uses for its supply of energy.

**Gluteus maximus**  The buttock muscle which you sit on. It is the largest muscle in the human body.

**Heart**  An organ within the chest which pumps blood through the arteries to all the body's living cells.

**Heart-beat**  The rhythmic beat made by the heart as it pumps blood around the body. If you put your hand near the centre of your chest, you can feel your own heart-beat.

**Hip replacement**  An operation carried out in hospital. Old worn-out hip joints are replaced by new artificial hip joints.

**Intestines**  Part of the digestive system.

**Lactic acid**  The substance made when cells do not have enough oxygen to burn glucose completely. A build-up of lactic acid is what makes the muscles go into cramp.

**Ligaments**  Strong, but slightly stretchy material which holds joints together, and stops bones moving too far in any one direction.

**Lungs**  Two organs in the chest which are used to breathe. When air is breathed into the lungs, the blood there absorbs oxygen from the air, and releases carbon dioxide to be breathed out.

**Membrane**  A thin layer of tissue which surrounds and protects different parts of the body.

**Minerals**  Chemicals, such as calcium, which the body needs to stay healthy.

**Muscle fibres**  Thin strands of muscle which are gathered together in bundles. Together, these bundles form muscle tissue.

**Nerves**  Pathways which carry messages between the brain and all parts of the body.

**Oxygen**  A gas which is needed by every living cell in the body to live and work. It is found in the air we breathe, along with several other gases.

**Phosphates**  One of the minerals needed to make bones strong and healthy.

**Proteins**  These substances are found in many different foods, including fish, meat, eggs and cheese. They are needed by the body to build new cells and to repair old and damaged ones.

**Skeletal muscles**  All the muscles which cover the skeleton to give your body its final shape.

**Smooth muscle**  One of three types of muscle in the human body. It is found in the digestive system and blood vessels.

**Spinal cord**  The bundle of nerves which runs down the backbone, protected by the vertebrae in the spine.

**Spongy bone**  A strong mesh of bone with spaces in between, found inside the hard outer layer of bones. These spaces are filled with bone marrow.

**Stirrup**  The smallest bone in the human body, found inside the middle part of the ear.

**Synovial fluid**  An oily protective liquid found inside some joints which helps the bones to move smoothly.

**Tendon**  Body tissue which attaches muscle to bone.

**Triceps**  One of the muscles in the upper arm. It contracts to straighten the arm.

**Veins**  Tubes which carry de-oxygenated blood (blood that has had its oxygen removed) back to the heart.

**Vertebrae**  The individual bones which are joined together to make your spine, or backbone.

**Virus**  An organism which can cause disease. If a virus gets inside the body, it invades and attacks the body's cells. White blood cells, made in the bone marrow, kill viruses and bacteria. Viruses are much smaller than bacteria.

**Vital organs**  All the important organs in your body, including your lungs, heart and liver, that are essential for life.

**X-ray**  A technique which is used in medicine to take pictures of the bones inside the body.

# Index